BSCS Science Tracks

AF904084

INVESTIGATING
Properties of Matter

Student Book

Third Edition

BSCS
SCIENCE LEARNING

Kendall Hunt
publishing company

Third Edition

BSCS Project Staff

Brooke Bourdélat-Parks, Director, Instructional Materials Development
Betty Stennett, Science Educator
Stacey Luce, Production Coordinator
Chris Moraine, Production Specialist

BSCS Administration

Richard J. Shavelson, Chair, Board of Directors
Dennis Schatz, Vice Chair, Board of Directors
Daniel C. Edelson, Executive Director
Robert Foulk, Director of Operations

Second Edition

Project Staff

April Gardner, Deborah Jordan, Project Directors
Barbara Perrin, Director of Publications
Dottie Watkins, Production Coordinator
Stacey Luce, Manuscript Specialist, Permissions
Pamela S. Warren, Assistant to Center for Curriculum Development Director
Judy L. Rasmussen, Assistant to Center for Research and Evaluation Co-Director
Dave Somers, Colorado Springs, Colorado, Editor

BSCS Administrative Staff

Carlo Parravano, Chair, Board of Directors
Rodger W. Bybee, Executive Director
Janet Carlson Powell, Associate Director
Pamela Van Scotter, Director, Center for Curriculum Development
Marcia M. Mitchell, Director of Finance

Reviewers

Linda Block-Gandy, Sharon R. Linn-Jeffrey, Julie L. Norris, Donnie M. Seibel
Ward's Natural Science, Safety Review

Cover image © Shutterstock.com

Kendall Hunt
publishing company

www.kendallhunt.com
Send all inquiries to:
4050 Westmark Drive
Dubuque, IA 52004-1840

ISBN 978-1-5249-2255-9

Published in the United States of America

First Edition

BSCS Development Team

Nancy M. Landes, Project Director and Author, 1996–1998
Gail C. Foster, Author
Colleen K. Steurer, Author
Vonna G. Pinney, Executive Assistant
Linda K. Ward, Senior Executive Assistant
Rodger W. Bybee, Principal Investigator, 1994–1995
Harold Pratt, Project Director, 1994–1996
Janet Chatlain Girard, Art Coordinator, 1994–1996

BSCS Administrative Staff

Timothy H. Goldsmith, Chair, Board of Directors
Joseph D. McInerney, Director
Michael J. Dougherty, Assistant Director
Lynda B. Micikas, Assistant Director
Larry Satkowiak, Chief Financial Officer

Contributors and Consultants

Randall K. Backe, BSCS, Colorado Springs, Colorado
Judy L. Capra, Wheat Ridge, Colorado, Freelance Writer
Michael J. Dougherty, BSCS, Colorado Springs, Colorado
B. Ellen Friedman, San Diego, California
Cathy Griswold, Lyons, Oregon
Jay Hackett, Greeley, Colorado
Debra A. Hannigan, Colorado Springs, Colorado, Contributing Author
David A. Hanych, BSCS, Colorado Springs, Colorado
Karen Hollweg, Washington, DC
Winston King, Bridgetown, Barbados
Paul Kuerbis, Colorado Springs, Colorado
Donald E. Maxwell, BSCS, Colorado Springs, Colorado
Brenda S. McCreight, Colorado Springs, Colorado, Contributing Author
Mary McMillan, Boulder, Colorado
Marge Melle, Littleton, Colorado, Freelance Writer
Lynda B. Micikas, BSCS, Colorado Springs, Colorado
Jean P. Milani, BSCS, Colorado Springs, Colorado
Renee Mitchell, Lakewood, Colorado, Freelance Writer
Carol D. Prekker, Broomfield, Colorado, Contributing Author
Patricia J. Smith, Tucson, Arizona, Contributing Author
Terry Spencer, Monterey, California, Contributing Author
Patti Thorn, Austin, Texas, Contributing Author
Bonnie Turnbull, Monument, Colorado, Freelance Writer
Terri B. Weber, Colorado Springs, Colorado
Carol A. Nelson Woller, Boulder, Colorado, Contributing Author

Field-Test Teachers and Coordinators, Levels 1–3

Joanne Allen, Grade 3, Westport Elementary School, Westport, Maine
Helene Auger, Westport School District, Westport, Maine
Sheila Dallas, Grade 2, Bethany School, Cincinnati, Ohio
Pat Dobosenski, Grade 3, Pembroke Elementary School, Troy, Michigan
Nina Finkel, Grade 1, Whitter Elementary School, Chicago, Illinois
Mary Elizabeth France, Grade 2, Westport Elementary School, Westport, Maine
Carolyn Gardner, Grade 3, Calhan Elementary School, Calhan, Colorado
Shelly Gordon, Grade 2, Bingham Farms Elementary School, Birmingham, Michigan
Darlene Grunert, Birmingham Public Schools, Birmingham, Michigan
Terry Heinecke, Grade 1, Edgerton Elementary School, Kalispell, Montana
Katherine Hickey, Grade 1, Irving Primary School, Highland Park, New Jersey
Jan Himmelspach, Grade 1, Grayson Elementary School, Waterford, Michigan
Elizabeth Lankes, Grade 3, Bethany School, Glendale, Ohio
Barbara O'Neal, Grade 1, Calhan Elementary School, Calhan, Colorado
Cheryl Pez, Grade 2, Bethany School, Cincinnati, Ohio
Rochelle Rubin, Waterford School District-IMC, Waterford, Michigan
Elizabeth A. Smith, Grade 3, Grayson Elementary School, Waterford, Michigan
Melanie W. Smith, Grade 2, Washington Elementary School, Raleigh, North Carolina
Janet Smith-James, Grade 3, Bartle School, Highland Park, New Jersey
Catherine Snyder, Highland Park School District, Highland Park, New Jersey
Ingrid Snyder, Grade 1, Waterford Village School, Waterford, Michigan
Lee Ann Van Horn, Wake County Public School System, Raleigh, North Carolina
Kathy Wright, Calhan Elementary School, Calhan, Colorado

Reviewers, Levels 1–3

James P. Barufaldi, University of Texas at Austin
Larry W. Esposito, University of Colorado at Boulder
Brenda S. Evans, Department of Education, Raleigh, North Carolina
Randy Gray, National Weather Service, Pueblo, Colorado
Judith Johnson, University of Central Florida, Orlando, Florida
Eric Leonard, The Colorado College, Colorado Springs, Colorado
Brownie Linder, Northern Arizona University, Flagstaff, Arizona
Jerry Ludwig, Fox Lane High School, Bedford, New York
Mike Madsen, KKTV, Channel 11, Colorado Springs, Colorado
Cherilynn A. Morrow, Space Science Institute, Boulder, Colorado
Kathleen Roth, Michigan State University, East Lansing, Michigan
Barbara W. Saigo, Saiwood Biology Resources, Montgomery, Alabama
Gail Shroyer, Kansas State University, Manhattan, Kansas
Carol Snell, University of Central Florida, Orlando, Florida
Joseph Stepans, University of Wyoming, Laramie, Wyoming
Richard Storey, The Colorado College, Colorado Springs, Colorado
Joan Tephly, Marycrest University, Iowa City, Iowa
Jack Wheatley, North Carolina State University, Raleigh, North Carolina

Contents

Doing Science to Learn Science

Introduction

You do different types of work in school. In this lesson, you will learn about work you will do in science. This lesson is called *Doing Science to Learn Science*. Maybe you have never thought about that phrase before.

Read these questions. Draw a picture for each question. Your drawing should show your thinking.

1. What does it mean to do science?
2. What is a scientist like?

How Do You Do Science?

Do you wonder? Do you ask questions? You may try to find answers to your questions. You may share your ideas with others. If so, you might be doing science! In this module, you will do science to learn science.

Look at the graphic on the next page. People who do science do these things. To do science, you do not have to be in school. You can do science in your backyard. You can do science in your home or on the playground. What things on the graphic interest you? In this module, you get to do these activities. They will help you learn science.

Doing Science

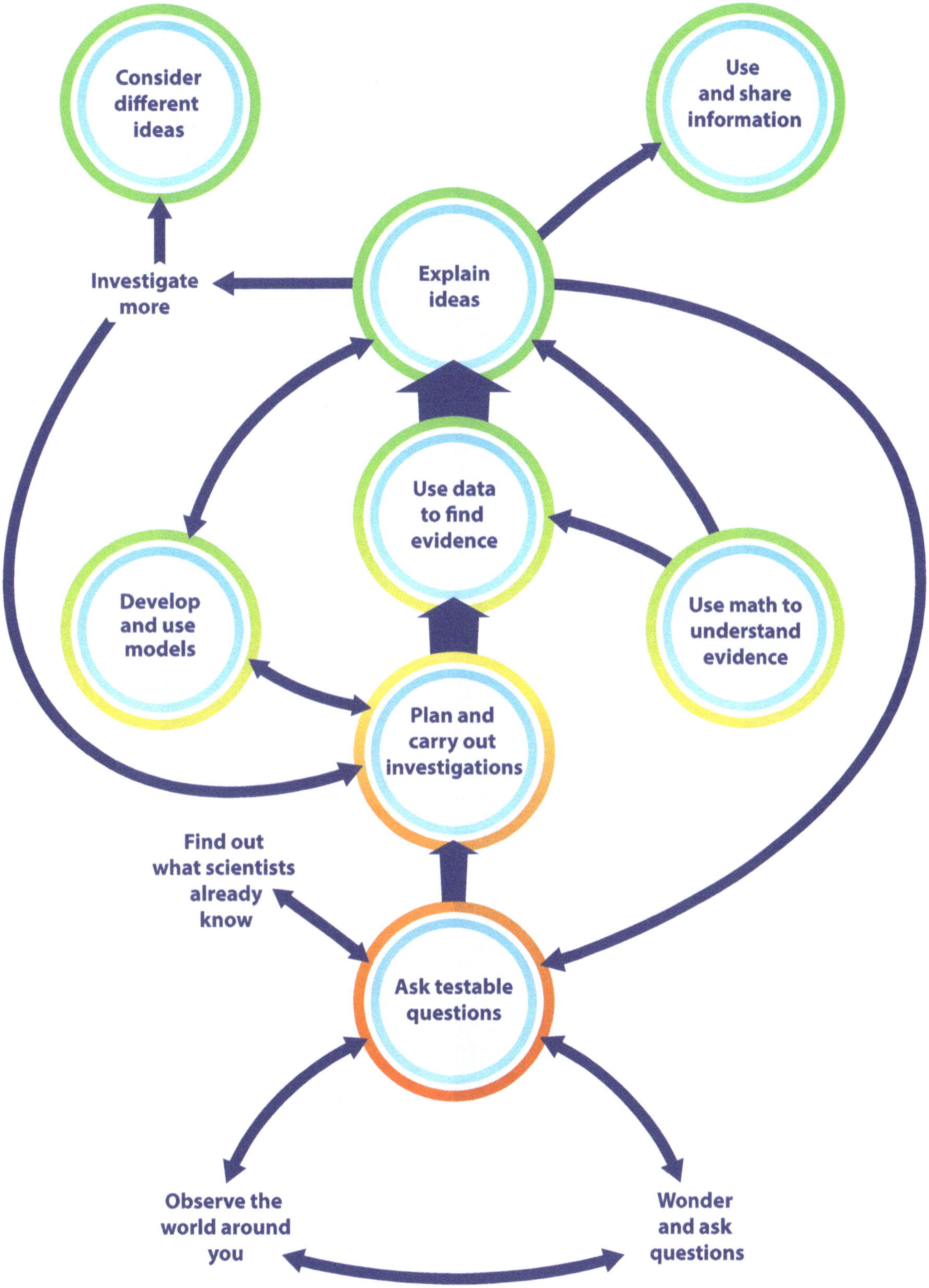

Ask Questions or Define Problems to Be Solved

Have you ever wondered about the world around you? Scientists wonder. They ask questions. They ask questions to find out more about the world. What questions do you have? Some questions you can investigate. That means you can answer the question by doing tests. These questions help scientists the most.

Scientists also define problems. They work to solve them. You may do this, too. Have you ever used something that did not work? You want to fix it or find someone to fix it. You have defined a problem that needs to be solved. This is part of doing science.

Develop and Use Models

When you do science, you may use a model. A model can help you do a lot of things in science. Sometimes it is hard to explain your ideas. When it is, often you draw a picture to help. Your picture is a type of model. A model will help you explain. A model can also help you understand things in the world that are too small to see. Or too big to see! Models may show things that move too slowly to notice. Or they move too quickly to notice. Maybe you will use a model as you investigate. Maybe you will use a model to collect data in your investigation.

Investigate

People do science to find answers to their questions. They observe. They use their senses. Their senses help them to observe. Imagine a cookie laying on the kitchen counter. Yum! Which senses would you use to observe the cookie?

Observing may include measuring. To measure, you must use a tool. Have you ever measured something? What tools did you use?

You want to find answers to your questions. So, you must do an investigation. To do this you will work with a team. It will be important to work together. You will plan together. You will follow your plan together. You will keep careful records. You might use or build a model. What you find out will help you answer your questions.

Use Data to Understand Evidence

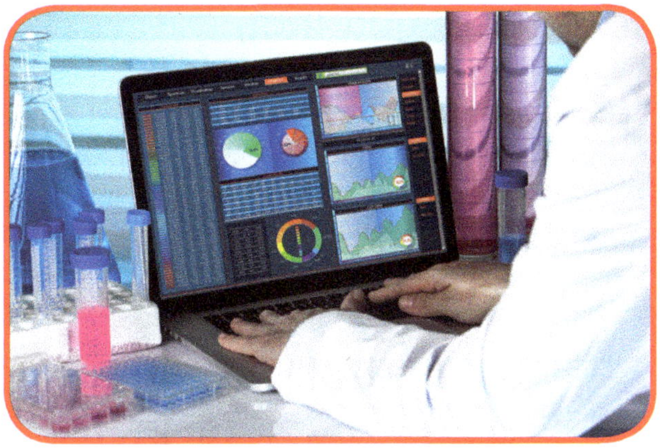

When you do science, you observe. You also measure. Your notebook is a place you write these things down. These observations and measurements are data. You may get data when you investigate. Scientists use data to answer their questions. You will, too! But, how do you use data?

Which color does the class like best?

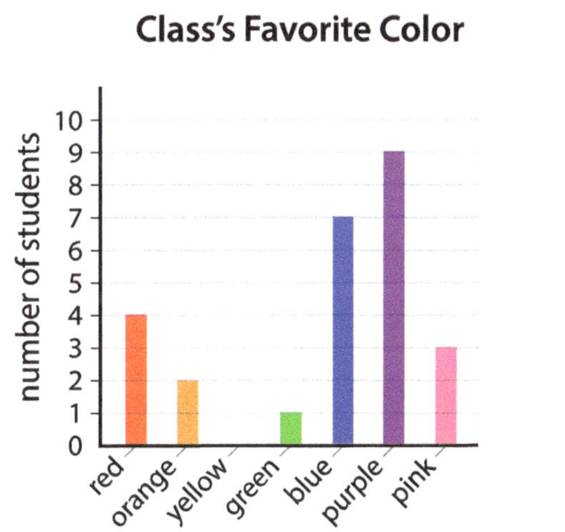

Data should be recorded. You might write down what you observe in a table. Perhaps you draw a picture of what you see. You might make a graph of what you measure. Your notebook is a good place to keep all your data.

You may see patterns in the data you record. What pattern do you see in the graph above?

Can you answer the question? You have used data to find evidence. Evidence is how you know something is true. You are doing science to learn science!

Use Math to Understand Data

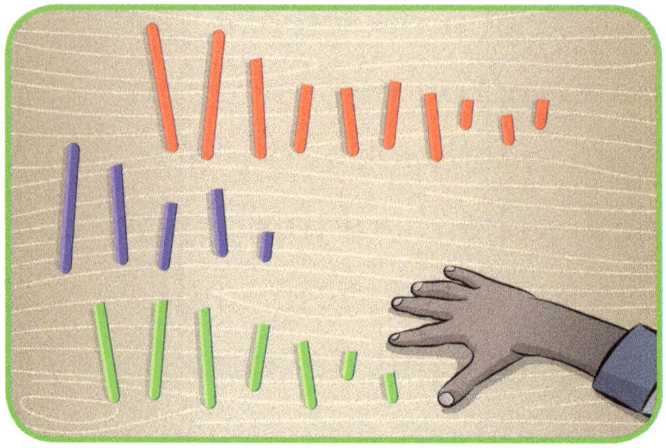

Sometimes you may use math in science class. Using math is important in science. Have you ever used math in science? Imagine you have a box of plastic sticks. There are different sizes of sticks. It may not be enough to describe the sticks as short or long. You need to know if the sticks are longer than another thing. Or are they shorter than the other thing? So, you rank the sticks by length. You put them in order from longest to shortest. That way you can compare their lengths. Ranking items is using math.

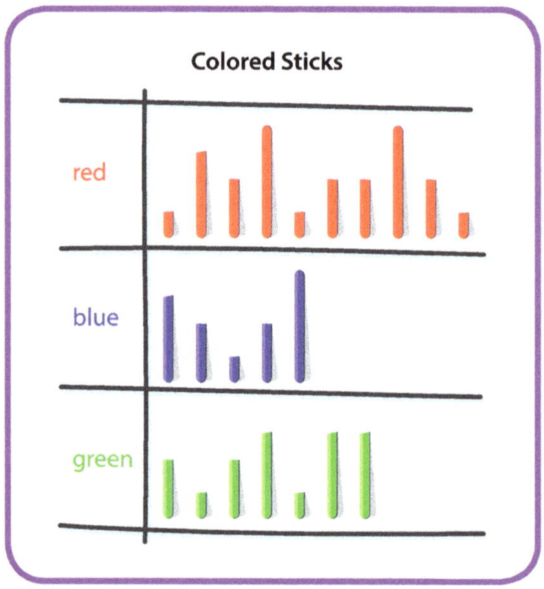

Now suppose some of your sticks are red. Some are blue. And some are green. Your team wants to know if there are more red, blue, or green sticks. This is a different question. You will use a different type of math to answer it. You can draw a pictograph. It will help you answer your question. Can you use the pictograph to answer the question? If so, you are using math to do science!

Explain Ideas

People who do science do not just answer their questions. They use their data to explain their answers. They may use a model to help them explain. Remember that a model can be a drawing. Models can also be something you build. You will use your data and models. They will help you explain the answers to your questions.

Consider Different Ideas

Arguing is part of doing science. But, arguing in science is not the same as arguing with your classmate. Argument in science means that you think about ideas that are different from yours. You consider which is best based on what you know. You listen. You ask questions. You share your ideas. You allow your ideas to grow and change.

When you argue in science, it is important to use evidence. Evidence helps you show why you think your ideas are right. You might already know some evidence. You can read or investigate to find out more.

You hear others talk about their ideas. You think about their evidence. Are they leaving any evidence out? Does the evidence make sense? Do you agree with their argument? Do you disagree? Why?

Use and Share Ideas

People who do science share what they know. They share what they find out. They share their ideas. They share their explanations. They may write or draw about their ideas. They talk about how their ideas are the same as others. They tell how their ideas are different. You will do this, too. You may read about science ideas. Then you may use these in your explanations. You share these ideas, too. You put the science ideas together with your ideas. This helps you make your point. Often, you use and share these ideas with your teammates. You will also use and share your ideas with your class.

Look back at the Doing Science graphic organizer on page 3. Can you find in the graphic all the things you read about? What do you want to do to learn science? Which part looks the most exciting?

Often you will do science in a team. Teamwork is important in science. As you work in a team, you will keep a record of your work. Record keeping is important in science. The next part you read will help you do your best as a team. It will also help you know how to keep good records.

Working Together to Do Science

People who do science often work in teams. They share ideas and tasks. You will work in teams, too. At times, you will work with people you know. Other times, you will work with people you do not know well. Teammates work together to do a task. Teammates help other teammates learn.

Team Skills

It is important to practice team skills. When you are doing science, practice these skills. When everyone on the team does, your team will do science the best!

Team Skills

1. **Understand and do your team job.**
2. **Work with your team to complete your task.**
3. **Make sure everyone has a chance to share their ideas.**
4. **Let your ideas grow and change.**
5. **Ask for and give help.**

---

1. Understand and do your team job.

When you work as a team, each teammate will do one job that helps the team. Those jobs are Materials Manager, Messenger, Tracker, and Time Tracker. Each job is important. You will have a chance to do all the jobs at some time. The job descriptions are on page 17.

2. Work with your team to complete your task.

Pay attention to your teammates. Work with them to do your task. Do not wander around the room.

3. Make sure everyone has a chance to share their ideas.

Everyone should get to share their ideas. That means everyone talks, but not at the same time! Your team will take turns sharing. A team is better when all ideas are heard.

4. Let your ideas grow and change.

Scientists are always learning new things. They learn new things from other scientists. They learn from investigations. They learn from reading about science ideas. When they learn more, their ideas change. You will learn new things, too.

When everyone shares, you hear new ideas. Those ideas may be better than your own. Or, your ideas may explain better. The good part about science is that you can change your ideas! You may change your ideas when you learn something new. That is part of doing science!

5. Ask for and give help.

At times, you may not understand what you need to do. You may not understand how to answer a question. When this happens, practice this team skill. Ask for and give help. Ask your teammates for help. Your teammates may be able to help you.

You could also help your teammates. A teammate may be quiet. This may be because he or she does not understand. You can help your teammate by explaining. Everyone should understand what they need to do. Then your team will work best.

Team Jobs

These descriptions will help you do each job.

Materials Manager

- gets supplies and return supplies
- tells teacher if something is missing or broken
- all teammates help clean up

Messenger

- asks another messenger for help
- asks teacher for help
- when the teacher visits the team, everyone can ask and answer questions

Time Tracker

- watches the timer
- makes sure the team finishes on time
- asks the teacher for more time if needed

Tracker

- tracks team progress
- makes sure the team does each step
- records the step number if the team has to stop

Making and Using a Science Notebook

Keeping track of your work in science is important. You will make observations. You will record data. You will draw pictures. Doing these things will help you learn science. But you have to be able to find what you have recorded. You need a good place to keep all your work. Keeping all your work in a notebook will help you.

What goes in a science notebook? Read on to find out.

Setting Up Your Notebook for the First Time

Make Your Notebook Your Own

Write your name on your notebook. You may want to decorate the outside of your notebook. You can put stickers on the cover. Draw a picture. Decorate your notebook any way you like. Make it your own!

Page Numbers

Number the pages of your notebook. Your teacher will tell you how to do this. This will help you find what you have written in your notebook.

Table of Contents

Write "Table of Contents" at the top of the first page. Leave a few pages blank at the front of your notebook. These pages are for the table of contents.

What to Include for Each Lesson

Lesson Names

Write down the name of the lesson in the Table of Contents. Find a clean page in your notebook. Write the name of the lesson on the top line. You can find this name in your student book. Find the page number at the bottom of the page. Write this page number in the Table of Contents.

Date

Record the date in your notebook. Do this each time you are in science class. Record the date even if you are not starting a new lesson. This will help you find your work later.

Other Things to Include in Your Notebook

Data

Draw pictures. Make graphs. Write down things you notice. These are all kinds of data. There are other kinds of data, too. Make sure you use labels when you need to. Your student book will help you know what to write down.

Explanations

Part of doing science is explaining your ideas. You may do this in your notebook. Your explanations will change as you learn more. Be sure to show in your notebook how they have changed. One way to do this is to use a different color. Write your new ideas in a different color. This way you and your teacher can see how your ideas have changed.

Handouts

Your teacher may give you a handout. Put these in your notebook. This keeps all your work together. Your teacher will tell you how to put the handouts in your notebook. This picture 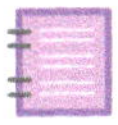 will remind you when to write in your notebook. You can record things that happen at other times, too.

Doing Science Safely

 Doing science is interesting and fun. You may work with things that can harm you or your clothes. Look for the caution icon. And then pay attention to the safety hints. Always follow the directions!

Digging In!

You have learned a little about doing science. You have begun to think about what it means to do science to learn science. You read about team skills and team jobs. You learned about notes to take in your notebook. You know what sign to look for to make sure that you work safely. Now it is time to dig in and begin doing science!

Chapter 1

A New Home for Dodger

A New Home for Dodger

Engage

I Spy!

Key Idea

- Some words are better than others to describe objects.

Linking Question:
Why do the same words describe different objects?

Big Ideas

- Objects can be sorted and compared by their properties.
- Materials can be tested to see which one is best for a job.

Explore

Building Materials

Key Idea

- Objects can be sorted and compared.

Linking Question:
What makes some things better for building a doghouse?

Explain

Materials to Use

Key Idea

- The properties of an object make it a good choice for some jobs.

Evaluate

Evidence of My Learning

Key Idea

- Properties are used to sort materials and decide on their best uses.

Linking Question:
How can we use what we know to choose the best stuffing for a dog bed?

Elaborate

Evidence for Keeping Dry

Key Idea

- Doing a test on an object gives us evidence to know if it is good for a job.

Linking Question:
How can I know which materials are best for a doghouse?

A New Home for Dodger

Have you ever hugged a dog? You know they are often soft and silky. You may have heard the dog bark. You can see that the dog is black with a red collar. You can smell his breath when he licks you.

Dogs are fun to play with. They like to run and chase a ball. Dogs like to sleep, too.

What kind of balls are best for dogs?
What bed is good for a dog?

As you work, you will learn how to find the best things for a dog named Dodger.

I Spy!

Do you have a pet? Many pets spend part of the day outside. Katy and Davis have a new dog. His name is Dodger. Dodger will stay outside part of the day. It rains a lot where Dodger lives, and he gets wet when he is outside. He needs some help from you and your class. Davis and Katy will help, too! Before you start, see if you can find some things to use.

Team Task

Work together to play the I Spy game with your class.

Team Jobs

Materials Manager

Messenger

Supplies

For the team

- 1 bag of items from your teacher

For each person on your team

- Science notebook
- Pencil

Steps to Take

1. Listen as your teacher reads part of the story, *A New Home for Dodger.*
2. Set up your science notebook. Start on a clean page. Put the title of this chapter on the first line. Put the name of the activity on the next line.

Go to next page.

Steps to Take continued

3. Read about your job. Take turns in your team and read the tasks aloud.

Team Jobs

Materials Manager

- gets supplies and returns supplies
- tells teacher if something is missing or broken
- all teammates help clean up

Messenger

- asks another messenger for help
- asks teacher for help
- when the teacher visits the team, everyone can ask and answer questions

4. Katy and Davis want to find things to use for Dodger's house. You can help them. Look in the bag that the Materials Manager gives you.

5. Listen as your teacher tells you about the game, I Spy. Make sure you know the rules.

Rules of I Spy

Student

- Think of an object in the bag.
- Think of a word that describes the object.
- Say the sentence: I spy something that is ________.
- Fill in the blank with your word.

Team

- Guess the object.
- Say a different word that describes the object.

Go to next page.

Steps to Take continued

6. Play the game, I Spy, with your team. Think about each thing you spy. Ask yourself, "Can I use this for a doghouse?"

7. Listen as your teacher leads a class discussion. Be sure to add your ideas.

Think About It

Answer these questions in your science notebook or on a handout.

1. Think of an object that you can see in the classroom. Try to find something that would work for Dodger's doghouse.

    a. Draw a box like that on the next page. Draw the object in the left side of the box.

    b. Write the name of your object in the blank under the box.

    c. Write three words to describe your object.

My Object	Words to Describe My Object
	1. ____________ 2. ____________ 3. ____________

My object is ______________________.

2. Your teacher made a list on the board. Which list helped you guess the object? Why?

3. How well did your team work together? Draw one of these symbols.

✓ We worked well together!

– We worked OK together.

✕ We did not work well together.

Explore

Building Materials

Katy and Davis found many things for the doghouse. They are about to put what they found in groups. There are many ways to sort things. You will work with some of the same things as Katy and Davis. You also may work with new things. Sorting will help you know if you have what you need for Dodger!

Team Task

Work with your team to sort items.

Team Jobs

- Materials Manager
- Messenger

Supplies

For each person on your team

- Science notebook
- Pencil

For each team

- 1 bag of items from your teacher

Steps to Take

1. Set up your notebook for a new activity. Write the title on the top line of a new page.
2. Review the team jobs. Make sure everyone on the team knows what to do.

Go to next page.

Steps to Take continued

Team Jobs

Materials Manager

- gets supplies and returns supplies
- tells teacher if something is missing or broken
- all teammates help clean up

Messenger

- asks another messenger for help
- asks teacher for help
- when the teacher visits the team, everyone can ask and answer questions

3. Send the Materials Manager to get a bag of items. Look at the materials that the Materials Manager has taken out of the bag. Spread them out on your table. Take turns with your teammates describing each item. Try to use only one word when you describe.
4. Look at the card that was in your bag. This card tells you how to sort your materials.

5. Katy and Davis are sorting the things they found. Sort your materials the way that the card says. Turn the card face down on your table when you have finished sorting. No one should be able to see your card.

6. Follow your teacher's directions to move to a different table. Do not touch the card on the new table. Think about how the materials are alike in each group. Can you guess how the materials are sorted?

7. Try to guess what is on the card without looking. Share your ideas with your team. You can share your ideas by saying the following:

   I think all the things in the first group are alike because they are ________________.

   I think all the things in the second group are alike because they are__________________.

   I think all the things in the third group are alike because they are____________________.

   Make sure everyone in your team shares an idea. When your teacher says it is OK, look at the card.

Go to next page.

Steps to Take continued

8. Go back to your desk. Get all the craft sticks from the Materials Manager. Make at least three groups. You choose how you will sort the sticks.

9. Now sort the craft sticks again. Make at least three new groups and sort them a different way.

10. Write in your notebook the two ways you sorted the craft sticks.

Think About It

Answer these questions in your science notebook or on a handout.

1. Name one thing from your bag that would be good for a doghouse.

   A ________________ would be good for a doghouse

   because it ______________________________________

   ____.

2. Draw a picture of three craft sticks that are different lengths. Draw them from shortest to longest.

   Shortest ————————————→ Longest

3. What Team Job did you do best? What did you do that helped the team?

Materials to Use

Katy and Davis want to build a doghouse. Their parents will have to help. They want to make sure the doghouse is big enough for Dodger. What can they use to build the doghouse? What are the best things to use? You will help Katy and Davis find the right things to build a doghouse that is big enough for Dodger.

Team Task

Choose items that will be good for a doghouse.

Team Jobs

- Materials Manager
- Messenger

Supplies

For the team

- 1 bag of items from your teacher
- Glue stick
- Paper
- Craft sticks

For each person on your team

- Science notebook
- Pencil

Steps to Take

1. Make sure you have written the title for this activity in your notebook. Use a new page.

Go to next page.

Steps to Take continued

2. Tell your teammates what your Team Job is. Tell them what you are supposed to do.
3. Take turns in your team reading about properties. Each person on your team should take turns reading aloud one sentence at a time.

Properties

Sorting things is fun! Sometimes you sort things by color. Other times you sort by length. Sometimes you get to make up your own rules for sorting. You have been using properties to sort things into groups. A **property** describes something. You may have sorted the craft sticks by color. You put all the same color in a group. If a craft stick had a different color, it went in a different group. Color is a property. You can describe a craft stick by its color.

Size is also a property. You sorted the craft sticks by size, too. Size may help you find the best things to use for the doghouse. What properties will be best? Do things need to be hard or soft? What size do they need to be? It is time to find out!

Go to next page.

Steps to Take continued

4. Do you want to help Davis and Katy build a doghouse? To help, find the best things to build a wall in the doghouse. To do that, you have to know what you have. The Materials Manager will spread the items on your table so you can see.

5. Work with your team to sort your items into 2 groups. Think about what you have that would be good for a wall of the doghouse. Use these 2 groups to sort.

Things that are …

GOOD
for a doghouse
wall

NOT
GOOD
for a doghouse
wall

6. Be ready to tell why you sorted the things the way you did.

7. You had craft sticks in one of your groups. Talk with your team about why craft sticks would be good for a doghouse. Take turns talking in your team. Everyone name one property of craft sticks that would make them a good choice.

8. There were different sizes of craft sticks. You need to choose the best size for Dodger's doghouse. You will need to test them to see if they are the right size. But your test needs to be fair. Read *A Fair Test* (on the next page) to learn how to be fair when testing size.

Go to next page.

Steps to Take continued

A Fair Test

Have you ever tried to see if you were taller than your friend? You stand next to each other. Someone else looks at the top of your heads to see who is taller. What if you were standing on your tip toes? Would that be fair? Not if your friend's feet were flat on the floor. You and your friend both have to stand with your feet flat on the floor. That way the test is fair.

You need to find the craft sticks that are the best size for Dodger's doghouse. You will help build a doghouse for Dodger. You will pretend Dodger is the size of the picture you colored. How big does your doghouse need to be? Do you have the right size craft sticks? You will have to measure your picture of Dodger. Where on Dodger will you start measuring? Where will you stop? Everyone will have to do it the same. That way the test will be fair.

9. Work with your classmates and your teacher to measure Dodger. Be sure to be fair!

10. Which picture shows the best doghouse for Dodger? Talk with your teammates about your ideas.

11. Now choose the best size of craft stick for Dodger's house. Follow your teacher's directions to build your part of the doghouse.

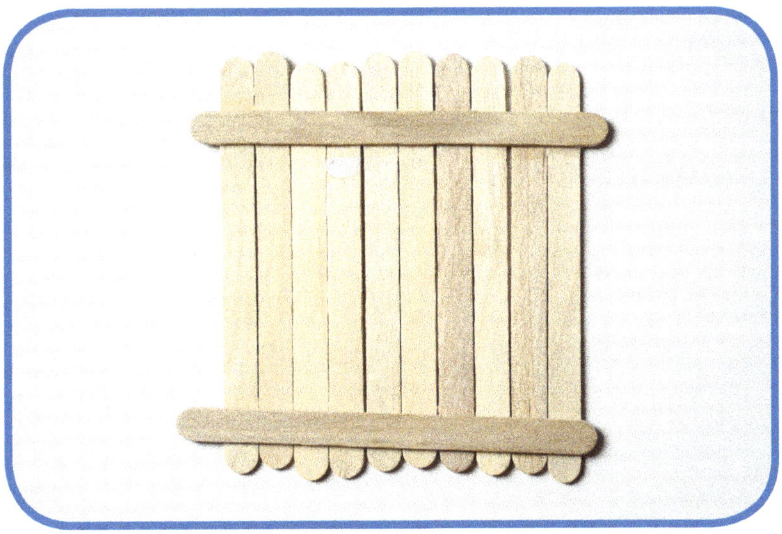

Think About It

Answer these questions in your science notebook or on a handout.

1. Do you think you built the best wall that you could for Dodger's house?

   If you answered "yes", why was it the best?

   If you answered "no", what could you do differently next time?

2. You just built part of a doghouse with craft sticks. What else could you build with craft sticks?

   Draw a picture of what you would build.

3. Sara had built a house out of blocks. Now she wants to build a car. But she only has a few blocks that she has not used. What can she do so she has enough blocks to build a car?

Evidence for Keeping Dry

You have learned about a fair test. A fair test helped you find the right size wood for Dodger's house. Now it is time to do a different fair test. Katy and Davis need your help! You need to plan a fair test to find the best fabric. The best fabric for this job is one that is waterproof. Using fabric that is waterproof will keep Dodger dry. Being waterproof is the *property* you are testing for. You will use your plan to test fabrics. So, do your best!

Team Tasks

Plan for a fair test of pieces of fabric.

Carry out the plan to get evidence for the best fabric.

Team Jobs

Messenger and Tracker

Materials Manager

Time Tracker

Supplies

For the team

- 1 pie tin
- 3 different types of fabric
- Water

- 2 large plastic cups
- 1 strainer
- Medium syringe

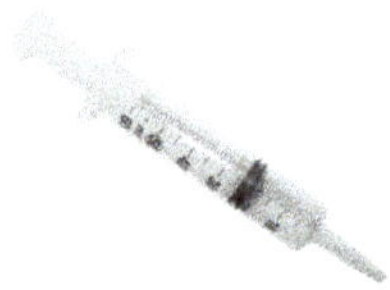

For each person on your team

- Science notebook
- Pencil

Steps to Take

1. Set up your notebook for a new activity.

2. You have supplies at your desk. Talk with your teammates about how you think you could use each item in your test. Your teacher will describe how to use them.

3. Your tests will give you evidence about how waterproof each fabric is. You will be trying to answer the question, *Which fabric is the best for our doghouse?* Read *Finding Evidence* (on the next page) to learn about evidence.

Go to next page.

Finding Evidence

Have you ever heard the word "evidence"? Maybe you heard the word on TV. **Evidence** is how you know something is true. Think about this example. You see your mom in the kitchen. She is mixing something in a bowl. A little later, you hear the oven make a ding. You smell something yummy. You have evidence that your mom is baking cookies. The evidence was what you saw, heard, and smelled. This evidence helped you to know that your mom was baking cookies.

You will test different pieces of fabric. You want to find out if any of them are waterproof. Your test may show that water will not go through the fabric. That is evidence that the fabric is waterproof. Your tests may also show that some fabrics are *not* waterproof. This would be like the roof on Dodger's doghouse. Dodger's roof was not waterproof. What is the evidence that tells you Dodger's roof is not waterproof?

4. Picking the best fabric to keep Dodger dry is important. Plan with your team how you will test the fabric. Your test should

- be fair!
- test three pieces of fabric.
- give evidence that the fabric is or is not waterproof.
- use only the materials at your desk.
- include recording evidence on the handout, *Waterproof Evidence.*

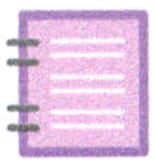

5. Make sure your team agrees on the steps. Then write the steps in your notebook. Everyone should do this part. Number your steps 1., 2., 3., and so on.

6. Take turns reading each of your steps out loud. Change your step if you need to so it makes sense.

Go to next page.

Steps to Take continued

7. Now it is time to use your plan. You must find the best fabric for Dodger's doghouse. Work with your team to set up your test.

8. Follow your plan to test each fabric. Test the fabrics one at a time. Send the Materials Manager to get water in the syringe. Don't forget to do these things:

   ✓ Test all three fabrics in the same way.
   ✓ Test fabrics one at a time.
   ✓ Record your evidence on your handout.
   ✓ Switch team jobs with each new test.

9. Work with your team to rank your fabrics. Rank them from the least waterproof to the most waterproof. In your notebook, record the fabric numbers in order of rank.

Think About It

Answer these questions in your science notebook or on a handout.

1. Name one step you did that made your test fair. How does it make the test fair?
2. Which piece of fabric do you think will be the best to keep the doghouse dry?
3. What evidence do you have that the fabric you chose is the best?
4. Which team job did you do best? What did you do that helped the team?

Evaluate

Evidence of My Learning

Dodger has a new home. It is dry inside his doghouse when it rains. His doghouse is roomy. And best of all, you helped build it!

Now your teacher needs some evidence. Now is the time to show yourself and your teacher what you have learned. Davis and Katy need your help, too. You need to find something to use for Dodger's bed. You will use what you know about sorting, properties, and evidence. Do your best work!

Supplies

For each student

- A bag of items
- Sorting mat
- Crayons or colored pencils
- Glue stick
- Science notebook
- Pencil

Steps to Take

1. What do you have in your bag? Spread the items on your desk to see.

Go to next page.

Steps to Take continued

2. Use your sorting mat to sort the objects by their properties. You decide what the groups are. Make at least 3 groups. All the items you put in a group should have similar properties.

My Sorting Mat

Group 1 | Group 2 | Group 3 | Group 4

Property of Group 1	Property of Group 2	Property of Group 3	Property of Group 4

3. Raise your hand when you are finished sorting. When your teacher sees, get your teacher to check your work.

4. While you are waiting for your teacher, write or draw one property of each group. Write it in the boxes below the groups. This is on your sorting mat.

Go to next page.

Steps to Take continued

5. Turn your sorting mat over. Sort your objects again. This time make two groups. One group has things that would NOT be as good for Dodger's bed. Place these objects off the mat. The other group has things that would be great for Dodger's bed.

Properties **good** for Dodger's bed:

Soft → Softest

6. Put all the items that would NOT make good stuffing back into the bag. Now look at everything you have that will make good stuffing. Rank these things from soft to softest. This arrow is on your sorting mat.

Soft → Softest

7. Choose the best item for stuffing Dodger's bed. Glue this at the top of the picture of Dodger's bed on your sorting mat. Glue the rest of the items you ranked on the arrow on the sorting mat in order from soft to softest.

8. Your choice for stuffing has properties. What properties make it a good choice? Write the properties below the picture.

Think About It

Answer these questions in your science notebook or on a handout.

1. What was your favorite part of learning about properties?
2. Look at the last picture in Dodger's story. Name two pieces of evidence that tell you it is getting colder.

3. Name two properties of snow.

Back and Forth

Back and Forth

Engage

Changing Properties

Key Idea

- Changing the temperature of a material can change its properties.

Big Ideas

- Changing the temperature can change some of the properties of matter.
- Solids and liquids have different properties.
- Some solids and liquids change back and forth because of changes in their temperature.

Linking Question:
What are the properties of solids and liquids?

Explore

Shape Shifters

Key Idea

- Some materials keep their shape, while some materials change shape.

Linking Question:
What causes the changes between solids and liquids?

Explain

Solids and Liquids

Part 1: What Are Solids and Liquids?

Key Idea

- Solids and liquids have different properties that can change with heat.

Part 2: Adding Heat

Key Idea

- Adding heat can cause changes.

Evaluate

A Chocolate Mess!

Key Idea

- Adding heat causes changes to some materials. Some of those changes are reversible and some are not.

Linking Question:
What evidence will help explain changes caused by heat?

Big Ideas continued

- Heating matter can cause some changes that are reversible and some that are not.
- Evidence can support a claim about reversible changes.

Elaborate

Switch-A-Roo

Key Idea

- Some changes caused by heat are reversible and some are not.

Linking Questions:
Part 1: What can cause a change from solid to liquid?
Part 2: Can these changes go back and forth?

Back and Forth

Dodger has a problem. He cannot drink his water. What happened? Can you use what you have learned about properties to help Dodger? You will have to learn more about properties and changes in properties to help Dodger.

You learned a lot about properties in the last chapter. The water in Dodger's bowl has turned to ice. The properties changed.

How did the properties change?
What caused the change?
Will it change back?

These are questions you may have about Dodger's water. In this chapter, you will find answers to these questions.

Engage

Changing Properties

Oh, no! Dodger cannot drink his water. What happened? Can you describe the properties of what is in Dodger's bowl? Are they the same as they were yesterday? Did they change?

Let's investigate to find out!

Team Task

Work with your team to

- describe properties of objects and
- notice changes in properties.

Team Jobs

Materials Manager

Messenger

Team Materials

- 1 handout, *Changing Properties*, for each team member
- Science notebooks
- Pencils
- 3 toothpicks
- Materials from your teacher

Steps to Take

1. Set up your notebook for a new chapter and new activity.
   Write these things:
      Chapter 2: Back and Forth
      Engage: Changing Properties
   Add today's date.

Go to next page.

Steps to Take continued

2. With your team, review the team jobs and what you are supposed to do.

   Look at the Team Jobs handout if you need help.

3. Share what you already know about properties. Do this with your team to share:

- Send the Materials Manager to get the first item.
- Look at the item with your team.
  - Use the toothpick to check to see if it is hard or soft.
- Discuss the properties of the item.

4. Draw a picture of the item on the handout your teacher has given you. List as many properties as you can.

5. Have the second Materials Manager return the item and get another one. Do Steps 3 and 4 with each new item.

6. Your teacher will do the next step. Watch carefully. Look for changes in properties of the item.

Think About It

Answer the following in your notebook or on the handout.

1. Your teacher used a hair dryer to change some of the items. What do you think caused the changes that you saw?

2. Draw a picture of the ice. Draw a picture of the water. Describe the shape of each.

ice

The shape is____________

______________________.

water

The shape is____________

______________________.

Explore

Shape Shifters

Dodger's water isn't wet anymore. His water is hard!

Let's look at Dodger's water before it got hard. You will compare it to the way it is now. You will get to look at other things, too.

How are the properties different? How are they the same? Do things other than water behave the same way? You will get to explore these ideas in this activity.

Team Task

Work with your team to

- explore the property of shape and
- find evidence to answer a question.

Team Materials

- Science notebooks
- Pencils
- Materials from teacher

Team Jobs

Messenger

Materials Manager

Tracker

Steps to Take

1. Set up your notebook for a new activity. Remember to write the activity title and date.
2. Take turns with your teammates to review each of the team jobs. You will get to do more than one job.

3. Your team's job is to look for evidence to answer a question. The question is:

What happens to the shape when I move something from place to place?

You learned about evidence in the last chapter. ***Evidence*** *is how you know something is true. Evidence also helps you answer a question. How do you know the shape of the water either changed or stayed the same?*

4. You will be looking for the shape. You will notice if the shape changes. Shape is a property of each item. Follow the process shown on the next page with your team to test two items.

Go to next page.

Steps to Take continued

Testing Two Items

a. Send the Materials Manager to get the first item. Everyone will do water first.

b. Write the name of the item on your handout. For this first test, write **water** on the handout.

c. Draw a picture of the water in container #1. Be sure to draw the shape of only the water.

d. Very carefully, move or pour the water into container #2.

e. Notice if the shape changed. Circle either **changes** or **stays the same** on your handout.

f. Draw a picture of the water in container #2. Be sure to draw the shape of only the water.

5. You have noticed the shape of water and if it changes when it is moved. Do other things behave the same way? Let's do another test to find out.

- Switch jobs. Send the new Materials Manager to get something different from your teacher.
- Repeat Step 4 with the new item.
- You will try to answer the same question as before.

What happens to the shape when I move something from place to place?

6. Not everyone tested the same item. Now it is time to learn from your class. Be ready to share your ideas in a class discussion. Listen for patterns as you discuss.

Think About It

Answer the following in your notebook or on the handout.

1. Ryder accidently spilled a glass of milk on the floor. What shape would the milk be? Use words and pictures to describe what you would see.
2. Some things changed shape and some things didn't. Name or draw 3 things that change shape when moved. Name or draw 3 things that do not change shape when moved. Try to name or draw something different from what you tested.

Things that do not change shape	Things that do change shape
1.	1.
2.	2.
3.	3.

Explain

Solids and Liquids

Dodger's water changed overnight. The water started as something Dodger could drink. The next morning, he could not drink it. What caused this change? Do you have ideas?

You may have seen ice cream change. Do you know what those changes are called?

In this activity, you will learn some new words. These new words will help you talk and write about these changes.

Part 1: What Are Solids and Liquids?

Team Task

Work together to learn about solids and liquids.

Materials

- Science notebooks
- Pencils

Steps to Take

1. Set up your notebook for a new activity.
2. Your teacher will make a 2-column chart on the board. Share your ideas as your teacher fills in the chart. Copy this chart in your notebook.
3. Read about liquids with your team. Each person on your team should take turns reading aloud one sentence at a time.

Liquids

Liquids are all around you.
You see them every day.
The things you drink are liquid.

But you should not drink all liquids!

Where else do you see liquids?

You can call something a liquid if it changes shape when moved. A liquid will change shape when you pour it into a different container.

This liquid starts out with the shape of the pitcher.

Go to next page.

Steps to Take continued

After pouring, the liquid is the shape of the glass.

Liquids take the shape of the container they are in.

Liquids often feel wet. Feeling wet is evidence that something is liquid. If you can pour it, that is more evidence that it is liquid. And, if it changes shape when moved, that is even more evidence!

4. One of the columns in the class chart has examples of liquids. Which one is it? Work with your team to decide which one to label "liquid". When you are finished, work with your class to label the column.

5. The items that are not liquid have different properties. What are they? Talk with your team about your ideas.

6. Read about solids with your team. Each person on your team should take turns reading aloud one sentence at a time.

Solids

Solids are also all around you. You see solids every day.

Solids may feel hard.

Go to next page.

Or solids may feel soft.

But all solids keep their shape when you move them from place to place.

You cannot pour a solid like you can a liquid. A baseball is a solid. A pencil, a piece of ice, or a desk are solids. Can you think of more solids?

A solid can bend like a paper clip.

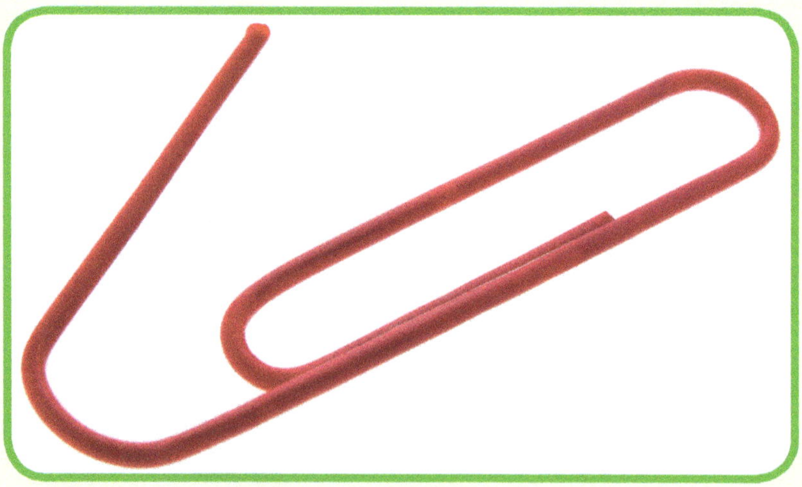

A solid can also break like a pencil.

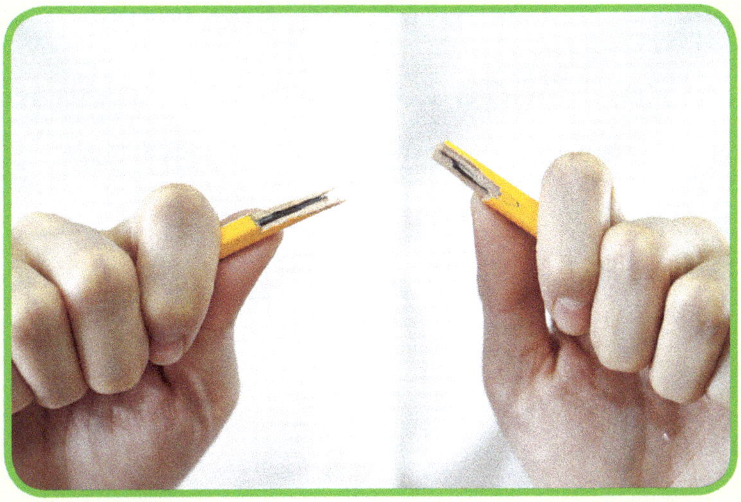

Both a paper clip and a pencil are solid. They do not change shape when you move them from place to place. That is evidence that they are solid.

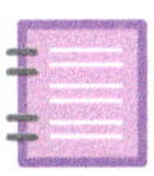

7. The class chart has one column that is not labeled. Do you know what the label should be? Add the label to your chart in your notebook.

8. Work with your class to do the next task. This will help you better understand solids and liquids.

Think About It

Answer the following in your notebook or on the handout.

1. Mr. Cantu gave Zac a cup with something in it. Zac listed the properties of what was in the cup. This is his list.

   Properties

   1. It pours.
   2. It is blue.
   3. It changes shape when I pour it.
   4. It feels wet.
   5. It smells good.

   Do you think what Zac has is a liquid or a solid? Choose one.

   Liquid **Solid**

2. List the properties that helped you answer question 1.
3. Which properties from question 1 did *not* help you answer question 1? List the properties that did not help you.

Part 2: Adding Heat

Now you know that water is a liquid. You also know that ice is a solid.

Dodger's water turned to a solid. We call solid water "ice". Can we change it back to a liquid? What causes the change?

Team Task

Work with your team to investigate if heat causes a change.

Team Jobs

- Messenger
- Materials Manager
- Tracker

Team Materials

- Science notebooks
- Pencils
- 1 *Adding Heat* handout for each team member
- Hair dryer
- Cups of materials

Steps to Take

1. Set up your notebook for a new activity.

2. Review your team jobs. Tell your team what you are supposed to do. Write your team job in your notebook.

3. Think about Dodger's water. It began as a liquid. Then overnight it changed to a solid. Do you think it will change back? Talk with your team about your ideas. Write your best ideas in your notebook.

4. Take turns in your team reading about heat. Each person on your team should take turns reading aloud one sentence at a time.

Heat

Heat is something you can feel. We feel heat when our bodies are cooler than the things around us.

You can feel heat coming from a warm cup of cocoa. This is because the cocoa is warmer than you are.

You can feel heat on a sunny summer day. The Sun is much warmer than you are, so you feel warm in the summer sunshine.

Go to next page.

Steps to Take continued

You can also feel the heat coming from a hair dryer. Which is warmer, you or the air from the hair dryer?

Heat is always moving. It moves from things that are warmer to things that are cooler. When you feel something warm, the heat is moving. That is why you feel the heat from the cocoa. It is moving from the warm cocoa to you!

A hair dryer can blow warm air. You will use the heat from a hair dryer to cause some changes. Adding heat can cause properties to change. Do the properties always change with heat? You will answer this question next.

5. Work with your class to prepare for your first test.
6. On your handout, draw a picture of the ice that is in the cup. Do not draw the cup. Draw your picture in the box labeled "ice". Circle either solid or liquid.

7. Do you think that heat from the hair dryer will change the properties of the material? If so, circle Yes on the handout. If you don't think heat will cause a change, circle No.

You just made a prediction. A ***prediction*** *is what you think will happen.*

8. You will do the next steps to test your prediction. Your test will help you answer the question, *Does adding heat cause a change in objects?*

Testing My Prediction

a. Hold the hair dryer near the cup. Hold it the same distance for every test. The Materials Manager should do this. Everyone will get to do this job during the activity.

Remember learning about a fair test. Holding the hair dryer the same distance every time helps make your test fair.

b. Turn on the hair dryer.

Do not touch the cup or the objects in the cup. These will get hot from the hair dryer.

Go to next page.

Steps to Take continued

c. Watch inside the cup for any changes. Make sure your teammates can see.

d. Leave the hair dryer on until the timer goes off.

e. Write down how the material changed. You can also draw a picture.

Some materials may not change. If they do not change after using the hair dryer, write "No change."

9. Switch roles and send the new Materials Manager to get the next set of materials. Make a prediction and test it by following Steps 7 and 8. Do these steps for each new thing you get.

10. Does adding heat cause a change? Talk about your ideas with your team.

11. The question you were trying to answer was, *Does adding heat cause a change*? Work with your teacher and your class to answer the question. The answer to a question is called a **claim**. You will be making a claim with your class.

Think About It

Answer the following in your notebook or on the handout.

1. We call the change from solid to liquid **melting**.

   Write "melting" on the correct arrow.

   Write "add heat" on the correct arrow.

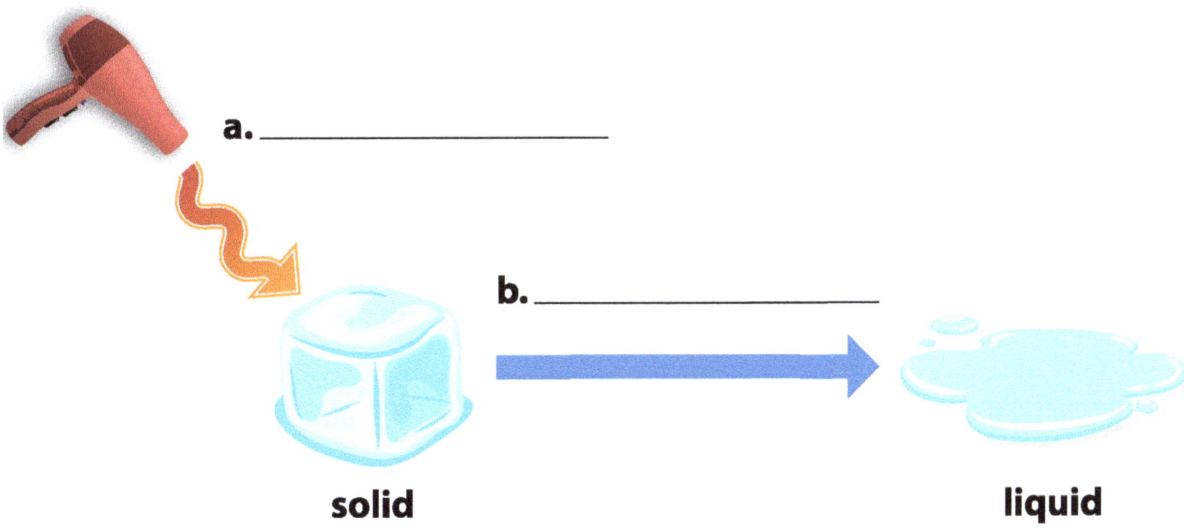

Go to next page.

Think About It continued

2. It was cold outside when Dodger's water turned to ice. Will it ever turn back to liquid water? How will that happen? Make another claim with evidence by completing the sentences below.

Your claim: Dodger's bowl of solid ice

_______________ change back to liquid water.
can / cannot

Your evidence: The ice will change to water when

_______________________________________.

I know this is true because ____________________

_______________________________________.

Switch-A-Roo

Good job! You found out that Dodger's water will not stay solid forever. In fact, Davis and Katy's dad didn't even have to do anything. Once the Sun came out, it warmed up quickly. The heat from the Sun melted the ice in Dodger's bowl. He had water to drink again!

Dodger's liquid water changed to solid ice. Then the solid ice changed back to liquid water. Water did a "switch-a-roo"! Do other things change back and forth like water? Let's find out!

Team Tasks

Work together in centers to find out if changes are reversible.

Team Jobs

- Messenger
- Materials Manager
- Time Tracker
- Tracker

Team Materials

- Science notebooks
- Pencils
- 1 handout, *Switch-A-Roo Centers*, for each team member

Steps to Take

1. Set up your notebook for a new activity.
2. Review the team jobs with your teammates.

3. You will be thinking about reversible changes in this activity. What are reversible changes? Read about reversible changes to find out. Take turns in your team reading each sentence.

Reversible Changes

Reversible changes are changes that you can switch back. Dodger's water changed to ice. You found out that adding heat would change it back to water. This change is reversible. Are all changes reversible? You will do tests to answer that question.

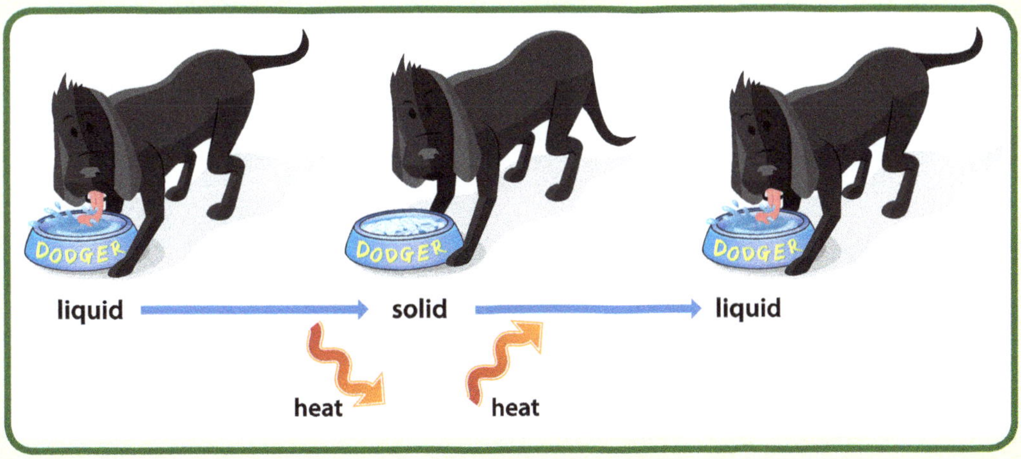

Go to next page.

Steps to Take continued

4. Listen as your teacher tells you about the centers. You will be trying to answer the question,

   Are all changes from heat reversible?

5. Move to your first center. Follow the directions at the center. Fill out the handout at each center.

Do steps 6 and 7 after you have finished all centers.

6. Your question is, *Are all changes from heat reversible?* Think about your answer. Then discuss the question with your team.

7. Make a claim that answers the question. Do this in your notebook.

   Your claim can start like either of these:

   Yes, all changes from heat are reversible. I know this because ______________________________.

   OR

   No, not all changes from heat are reversible. I know this because __________________________.

Think About It

Answer the following in your notebook or on the handout.

1. Make a 2-column chart like the one below. List the items from the centers in each column.

Reversible	Not reversible

2. Juan said that color changes are NOT reversible. Do you agree or disagree? What is your evidence?

3. Sara was trying to answer the question, *Is this change from heat reversible?* She made this claim: **I think that butter melting is reversible.**

   Sara listed these things as evidence:

   1. I like butter.
   2. I put melted butter on ice and it became solid.
   3. The butter is solid.
   4. The ice is cold.
   5. When heat is removed, the butter is solid.

   List the evidence that helps Sara answer her question.

   List the evidence that does not help Sara answer her question.

A Chocolate Mess!

You have learned that some changes are reversible. It is good that the change in Dodger's bowl was reversible! Now it is time for you to show what you have learned in this chapter.

You have learned a lot about changes.

You have learned what caused those changes.

And you have learned that some of those changes can be reversed.

And some changes cannot be reversed.

Steps to Take

1. Prepare your notebook for a new activity.
2. Read the story with your team.

Chocolate Changes

Katy and Davis went on a trip with their family. Their trip was during the summer. Their mom and dad got some snacks for the trip. One snack was Katy's favorite. Katy had three chocolate bars! Mom and Dad packed some drinks, too. They put the drinks in a cooler with ice.

Go to next page.

Steps to Take continued

The family drove through three towns before they stopped. Everyone was hungry. They stopped at a park to have a picnic.

After the picnic, Katy and Davis got back in the car. It was very hot in the car. Davis saw Katy's chocolate bars in the bag. They looked different.

3. Talk with your team about what happened to Katy's chocolate bars. Use the picture below to help you.

Go to next page.

Steps to Take continued

4. Work by yourself to write your ideas in your notebook. Here are some words you may want to use. You should not use all of them. Use only the ones that describe Katy's chocolate bars now.
   - solid
   - liquid
   - changes shape
   - does not change shape
   - hard
   - melted
   - frozen
5. Katy's chocolate bars were fine when they left home. Now they are a mess! Do you think you can change the chocolate bars back? Make a prediction. Write your prediction in your notebook. Use this sentence to help you.

   ______________________ change the chocolate bars back.
   (Yes, I can or No, I cannot)

6. If you think you can change the bars back, do this step. Think about what was in the family car. What would you use to change the chocolate bars back? List those in your notebook.

7. What would you do to test your prediction? Write steps in your notebook.

Think About It

Answer the following in your notebook or on the handout.

1. List 3 things you learned in this chapter.

2. What activities helped you learn these things?

3. Write 1 question you still have.

   I still wonder ______________________________.

Photo Credits

BSCS acknowledges Tahoka Elementary School, Tahoka, Texas, and the second-grade students of Mrs. D'Lynn Gill for their help with photographs for the Properties module. Pages 13, 16, 22, 71, 73, 78, 79, 85, 89, 90, 93, 98, 107).

Doing Science

Page 1: Lai Wei Suen / Shutterstock.com
Page 4: Scientist: Blend Images / Shutterstock.com; boy: Blend Images / Shutterstock.com
Page 5: Girl: Ilike / Shutterstock.com; scientist: Dragon Images / Shutterstock.com
Page 6: Scientists: Pressmaster / Shutterstock.com; girls: espies / Shutterstock.com
Page 7: angellodeco / Shutterstock.com
Page 9: Adults: OPOLJA / Shutterstock.com; children: Lorena Fernandez / Shutterstock.com
Page 10: Adults: Gorodenkoff / Shutterstock.com; children: InesBazdar / Shutterstock.com
Page 11: Children: Syda Productions / Shutterstock.com; adults: Matej Kastelic / Shutterstock.com
Page 17: Rob Marmion / Shutterstock.com

Chapter 1

Page 27: sujan Shetty / Shutterstock.com
Page 28: espies / Shutterstock.com
Page 29: Dmussman / Shutterstock.com
Page 34: My Life Graphic / Shutterstock.com
Page 40: Stephen Orsillo / Shutterstock.com
Page 44: Monkey Business Images / Shutterstock.com
Page 45: Stephen Orsillo / Shutterstock.com
Page 47: optimarc / Shutterstock.com
Page 48: ESB Professional / Shutterstock.com
Page 50: WOVE LOVE / Shutterstock.com
Page 51: sujan Shetty / Shutterstock.com
Page 52: Pie tin: lampang_naomark / Shutterstock.com; fabric: Jenny Solomon / Shutterstock.com; water: science photo / Shutterstock.com; plastic cups: Chimpinski / Shutterstock.com; strainer: topimages / Shutterstock.com; syringe: Ilona Koeleman / Shutterstock.com.
Page 53: Jenny Solomon / Shutterstock.com
Page 54: canovass / Shutterstock.com
Page 55: Monkey Business Images / Shutterstock.com
Page 58: Anna Hoychuk / Shutterstock.com
Page 61: Robert Kneschke / Shutterstock.com
Page 63: wavebreakmedia / Shutterstock.com

Chapter 2

Page 72: Monkey Business Images / Shutterstock.com
Page 77: Monkey Business Images / Shutterstock.com
Page 81: Blend Images / Shutterstock.com
Page 82: Boy: Firma V / Shutterstock.com; poison: F. JIMENEZ MECA / Shutterstock.com
Page 83: Pitcher and glass: science photo / Shutterstock.com; pitcher: Excentro / Shutterstock.com
Page 84: Cyril Hou / Shutterstock.com
Page 85: Rocks: Sakdinon Kadchiangsaen / Shutterstock.com
Page 86: Cotton: Alina Cardiae Photography / Shutterstock.com; paper clip: HomeStudio / Shutterstock.com
Page 87: Broken pencil: mnimage / Shutterstock.com; paper clips: Mr.Nattapong Udtha / Shutterstock.com; pencils: winuturn / Shutterstock.com
Page 91: Boy: all_about_people / Shutterstock.com; girl: XiXinXing / Shutterstock.com
Page 92: Damir Khabirov / Shutterstock.com
Page 100: Robert Kneschke / Shutterstock.com
Page 103: aastock / Shutterstock.com
Page 104: Batter: Tomophafan / Shutterstock.com; oven: Andrei Tudoran / Shutterstock.com; brownies: nikon84 / Shutterstock.com

www.ingramcontent.com/pod-product-compliance
Ingram Content Group UK Ltd.
Pitfield, Milton Keynes, MK11 3LW, UK
UKHW052013041026
15590UKWH00008B/2308

9 781524 9225